TAKE-OFF!

What is Weather?

SNOW

Andy Owen and
Miranda Ashwell

Heinemann
LIBRARY

First published in Great Britain by Heinemann Library,
Halley Court, Jordan Hill, Oxford OX2 8EJ,
a division of Reed Educational and Professional Publishing Ltd.
Heinemann is a registered trademark of Reed Educational & Professional Publishing Limited.

OXFORD MELBOURNE AUCKLAND
JOHANNESBURG BLANTYRE GABORONE
IBADAN PORTSMOUTH NH (USA) CHICAGO

Designed by David Oakley and Celia Floyd
Illustrations by Jeff Edwards
Originated by Dot Gradations, UK
Printed and bound in Hong Kong/China

04 03 02 01 00
10 9 8 7 6 5 4 3 2 1

ISBN 0 431 03783 3
This book is also available in hardback (ISBN 0 431 03778 7).

British Library Cataloguing in Publication Data

> Owen, Andy
> Snow. - (What is weather?) (Take-off!)
> 1. Snow - Juvenile literature
> I. Title II. Ashwell, Miranda
> 551.5'784

Acknowledgments
The Publishers would like to thank the following for permission to reproduce photographs:
B & C Alexander: pp5, 9, 16, A Hawthorne p22; Bruce Coleman Limited: p13, J Johnson p14, H Reinhard pp6, 20, K Taylor p4, M Taylor p22, S Widstrand pp18, 19; Robert Harding Picture Library: p23, N Blythe p28, L Burridge p7, Explorer p29, J Robinson p26; Oxford Scientific Films: C Monteath p8, B Osborne p15; Pictor International: p11; Planet Earth Pictures: S Nicholls p17; SIPA: Le Progres p27; Still Pictures: T Thomas p12; Tony Stone Images: J Stock p24.

Cover: Tony Stone/Hans Strand
Our thanks to Sue Graves for her advice and expertise in the preparation of this book.

For more information about Heinemann Library books, or to order, please telephone +44 (0)1865 888066, or send a fax to +44 (0)1865 314091. You can visit our website at www.heinemann.co.uk

> Any words appearing in the text in bold, **like this**, are explained in the Glossary.

Contents

What is snow?

Snow happens when it is very cold. Droplets of water in a cloud **freeze** when it is very cold. They freeze into ice crystals. The ice crystals all have six sides but make different patterns.

This snowflake is shown 20 times larger than its real size.

A heavy snowfall quickly covers the ground.

The ice crystals stick together. They get very heavy. Then they drop from the sky as snowflakes. On very cold days, snow can cover the ground.

The amount of snow that falls is ten times the amount of rain that would have fallen!

What is frost?

At dawn, trees, grass and plants are covered in water. This is because the air is full of tiny water droplets. These water droplets are called dew. The dew can **freeze** on cold nights and make a thin layer of ice on twigs and branches. This is called **frost**.

After a cold night, frost forms on twigs and branches.

Pond and rivers freeze over in the winter, but the ice can be very thin!

In winter, if it is very cold, rivers and ponds can be covered with a thin layer of ice. The ice may look strong, but it will break where it is thin. You should never walk across frozen ponds or rivers because the thin ice will break.

Snow on mountains

The air gets very cold as the ground becomes higher. Water droplets in the clouds **freeze** and fall as snow. Even in hot places, some mountains are always covered in ice and snow.

Even in summer, this mountain has a cap of snow on top of it.

snow cap

This glacier slowly slides down the mountain.

Over many years, the layers of snow on a mountain are slowly crushed together. The crushed layers of snow are turned into ice. The mass of ice slowly slides down the mountain in a **glacier**.

The coldest places

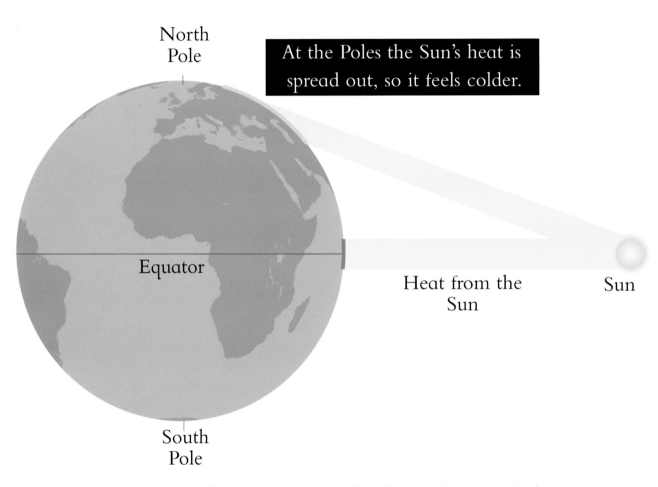

North
Pole

At the Poles the Sun's heat is
spread out, so it feels colder.

Equator

Heat from the
Sun

Sun

South
Pole

At the North and South **Poles**, the Sun's heat is weaker because the Sun is farther away and the rays are more spread out. The North and South Poles are the coldest places in the world.

Sunlight shining on snow is very bright. People have to be careful not to get snow blindness. They wear sunglasses to protect their eyes. Warm clothes are also needed near the Poles.

People wear warm clothes and sunglasses near the Poles.

Antarctica is not only the coldest place but also the windiest!

The North and South Poles

The South **Pole** is in **Antarctica**. The land at the South Pole is covered in snow and ice. Penguins live in Antarctica. They have thick feathers and a layer of fat on their bodies to keep them warm.

These penguins huddle together to keep warm.

This killer whale is swimming in the cold Arctic Ocean.

The North Pole is in the **Arctic**. There is no land here, just ice on the sea. Whales and seals swim beneath the thick ice.

Icebergs

At the **Poles**, there are huge **glaciers**. The glaciers move towards the sea. Warmer weather makes the ice melt. Huge chunks of ice break away from the glacier to become **icebergs**.

An iceberg breaks away from a glacier.

Icebergs are mountains of ice that float on the sea. Icebergs float because they are lighter than water. Nearly all of an iceberg is hidden under the water.

Icebergs are floating mountains of ice.

You can only see about 20% of an iceberg above the surface of the water!

Frozen ground

The land close to the **Poles** is covered in deep snow. It is very, very cold there. People use sledges, skis or snowmobiles to travel across the snow and ice.

Some people use snowmobiles, like this one, to travel across the thick snow and ice.

In the summer, the rivers fill up with water from the melted snow.

Land near the Poles is frozen hard. Only the top layer of the ground **thaws** during the short summer. Under this, there is a permanently frozen layer called permafrost.

17

Animals in the snow

Animals that live in cold, snowy places keep warm by having thick layers of fur. Some animals have white fur so they blend in with the snowy background and are hard to see.

This white wolf cannot be seen easily when she is hunting for food.

In winter, these hares have white coats to keep them safe from danger.

Some animals can change the colour of their coats to help them hide from danger. In winter, these hares grow thick, white coats of fur to hide them in the snow. But in the summer, when the snow melts, they grow brown coats to blend in with the land.

Plants in the snow

Mountain plants have special ways of growing in cold, snowy places. These flowers have thick, furry petals to keep out the cold. They also grow close to the ground to protect them from the wind.

The furry petals on these plants help to keep out the cold.

furry petal

Pine trees can grow higher up a mountain than any other tree.

These pine trees have special leaves, called needles, to protect them from the snow. The needles are thin and point downwards so the snow just slides off them.

Working in the snow

Snow and ice can **freeze** your fingers and toes and give you **frostbite**. People who work in cold, snowy places wear special clothes to protect them from frostbite.

This man is wearing special clothes to keep him warm in the snow.

goggles

hat

face mask

thick jacket

Reindeer are important to the Lapp people.

The Lapp people live among the wild reindeer.
They use thick, warm, reindeer skins to make boots
and clothes. Some Lapp people also use the skins to
make their homes.

Fun in the snow

Snowballs are best made out of wet snow. Wet snow is only just frozen so it is easy to roll into a soft, wet snowball.

Many children, and even adults, like to throw snowballs.

Snow and ice have smooth surfaces so they are very easy to slide on. A sledge moves very fast on slippery snow. Some people prefer to ski instead.

Children of all ages like to go sledging on the snow.

Blizzard!

A snowstorm is called a **blizzard**. Heavy falls of snow and high winds cause problems. People may be trapped when blizzards block roads. Snow ploughs are sent to clear the blocked roads.

Snow ploughs help to keep the roads clear of snow.

26

power-lines

The weight of the snow and ice has broken these power-lines.

The weight of a lot of snow and ice on power-lines can bring them down. Many homes will be without electricity if power-lines break. Without electricity, homes have no light or heat.

Snow accidents

Snow at the top of a mountain can suddenly fall down steep slopes. This is called an **avalanche**. An avalanche picks up more snow as it sweeps down the side of the mountain.

An avalanche sweeps down the side of a mountain.

These men take the rescued person off the mountain on a special sledge.

Avalanches are very dangerous. These men have rescued someone who was trapped in an avalanche. They had to work quickly to find this person.

It's amazing!

Antarctica is the coldest place in the world. The ice in Antarctica is thousands of years old. In some places it is 3000 metres thick.

Snow fell in the Kalahari **Desert** in Africa on 1 September 1981.

Inuit people used to build houses from blocks of snow.

The record for the most snow to fall during a 24-hour period is 190.5 centimetres at Silver Lake, Colorado, USA, on 14 April 1921.

Glossary

Antarctica	land at the southernmost part of the world, around the South Pole
Arctic	frozen area at the northernmost part of the world, around the North Pole
avalanche	a sudden fall of snow down the side of a mountain
blizzard	a storm of heavy snow
desert	a place that is very dry, and usually very hot
freeze	when water turns to ice
frost	droplets of water in the air, too small to see, that freeze and turn to ice
frostbite	injury to the body caused by freezing
glacier	a river of ice
iceberg	a mountain of ice floating in the sea
Inuit	people who live in the coldest parts of North America and Greenland
Poles	the North and South Poles are the coldest places on Earth. They are the furthest points from the Equator
thaw	a thaw happens when warmer weather makes the snow melt

Index